GLOBAL ISSUES

Series editors: Stephen Ha

PEOPLE AND TECHNOLOGY

Mary Fisher

*6 Studies
for individuals
or groups*

INTERVARSITY PRESS
DOWNERS GROVE, ILLINOIS 60515

InterVarsity Press is the book-publishing division of InterVarsity Christian Fellowship, a student movement active on campus at hundreds of universities, colleges and schools of nursing in the United States of America, and a member movement of the International Fellowship of Evangelical Students. For information about local and regional activities, write Public Relations Dept., InterVarsity Christian Fellowship, 6400 Schroeder Rd., P.O. Box 7895, Madison, WI 53707-7895.

All Scripture quotations, unless otherwise indicated, are from the Holy Bible, New International Version. Copyright © 1973, 1978, International Bible Society. Used by permission of Zondervan Bible Publishers.

Cover illustration: TransLight

ISBN 0-8308-4909-2

Printed in the United States of America

12	11	10	9	8	7	6	5	4	3	2	1
99	98	97	96	95	94	93	92	91	90		

Contents

Because humankind is made in the image of God, every person, regardless of race, religion, color, culture, class, sex or age, has an intrinsic dignity because of which he or she should be respected and served, not exploited. Here too we express penitence both for our neglect and for having sometimes regarded evangelism and social concern as mutually exclusive.

Although reconciliation with people is not reconciliation with God, nor is social action evangelism, nor is political liberation salvation, nevertheless we affirm that evangelism and sociopolitical involvement are both part of our Christian duty. For both are necessary expressions of our doctrines of God and humankind, our love for our neighbor and our obedience to Jesus Christ.

The message of salvation implies also a message of judgment upon every form of alienation, oppression and discrimination, and we should not be afraid to denounce evil and injustice wherever they exist.

—*Lausanne Covenant, Article Five.*

Welcome to Global Issues Bible Studies

With all the rapid and dramatic changes happening in our world today, it's easy to be overwhelmed and simply withdraw. But it need not be so for Christians! God has not only given us the mandate to love the world, he has given us the Holy Spirit and the community of fellowship to guide us and equip us in the ministry of love.

Ministering in the world can be threatening: It requires change in both our lifestyle and our thinking. We end up discovering that we need to cling closer to Jesus than ever before—and that becomes the great personal benefit of change. God's love for the world is the same deep love he has for you and me.

This study series is designed to help us understand what is going on in *the world*. Then it takes us to *the Word* to help us be faithful in our compassionate response. The series is firmly rooted in the evangelical tradition which calls for a personal saving relationship with Jesus Christ and a public lifestyle of discipleship that demonstrates the Word has truly come alive in us.

At the front of the guide is an excerpt from the Lausanne Covenant which we have found particularly helpful. We have developed this series in keeping with the spirit of the covenant, especially sections four and five. You may wish to refer to the Lausanne Covenant for further guidance as you form your own theology of evangelism and social concern.

In the words of the covenant's authors we place this challenge before you: "The salvation we claim should be transforming us in the totality of our personal and social responsibilities. Faith without works is dead."

Getting the Most from Global Issues Bible Studies

Global Issues Bible Studies are designed to be an exciting and challenging way to help us seek God's will for all of the world as it is found in Scripture. As we learn more about the world, we will learn more about ourselves as well.

How They Are Designed

Global Issues Bible Studies have a number of distinctive features. First, each guide has an introduction from the author which will help orient us to the subject at hand and the significant questions which the studies will deal with.

Second, the Bible study portion is inductive rather than deductive. In other words, the author will lead us to discover what the Bible says about a particular topic through a series of questions rather than simply telling us what he or she believes. Therefore, the studies are thought-provoking. They help us to think about the meaning of the passage so that we can truly understand what the biblical writer intended to say.

Third, the studies are personal. Global Issues Bible Studies are not just theoretical studies to be considered in private or discussed in a group. These studies will motivate us to action. They will expose us to the promises, assurances, exhortations and challenges of God's

Word. Through the study of Scripture, we will renew our minds so that we can be transformed by the Spirit of God.

Fourth, the guides include resource sections that will help you to act on the challenges Scripture has presented you with.

Fifth, these studies are versatile. They are designed for student, mission, neighborhood and/or church groups. They are also effective for individual study.

How They Are Put Together

Global Issues Bible Studies also have a distinctive format. Each study need take no more than forty-five minutes in a group setting or thirty minutes in personal study—unless you choose to take more time.

Each guide has six studies. If the guides are used in pairs, they can be used within a quarter system in a church and fit well in a semester or trimester system on a college campus.

The guides have a workbook format with space for writing responses to each question. This is ideal for personal study and allows group members to prepare in advance for the discussion. In addition the last question in each study offers suggestions and opportunity for personal response.

At the end of the guides are some notes for leaders. They describe how to lead a group discussion, give helpful tips on group dynamics and suggest ways to deal with problems which may arise during the discussion. With such helps, someone with little or no experience can lead an effective study.

Suggestions for Individual Study

1. As you begin the study, pray that God will help you understand and apply the passages to your life. Pray that he will show you what kinds of action he would have you take as a result of your time of study.

2. In your first session take time to read the introduction to the entire guide. This will orient you to the subject at hand and the author's goals for the studies.

3. Read the short introduction to the study.

4. Read and reread the suggested Bible passages to familiarize yourself with them.

5. A good modern translation of the Bible, rather than the King James Version or a paraphrase, will give you the most help. The New International Version, the New American Standard Bible and the Revised Standard Version are all recommended. The questions in this guide are based on the New International Version.

6. Use the space provided to respond to the questions. This will help you express your understanding of the passage clearly.

7. Look up the passages listed under *For Further Study* at the end of each study. This will help you to better understand the principles outlined in the main passages and give you an idea of how these themes are found throughout Scripture.

8. It might be good to have a Bible dictionary handy. Use it to look up any unfamiliar words, names or places.

9. Take time with the final question in each study to commit yourself to action and/or a change in attitude.

Suggestions for Group Study

1. Come to the study prepared. Follow the suggestions for individual study mentioned above. You will find that careful preparation will greatly enrich your time spent in group discussion.

2. Be willing to participate in the discussion. The leader of your group will not be lecturing. Instead, he or she will be encouraging the members of the group to discuss what they have learned. The leader will be asking the questions that are found in this guide.

3. Stick to the topic being discussed. Your answers should be based on the verses which are the focus of the discussion and not on outside authorities such as commentaries or speakers.

4. Be sensitive to the other members of the group. Listen attentively when they describe what they have learned. You may be surprised by their insights! When possible, link what you say to the comments of others. Also, be affirming whenever you can. This will encourage

some of the more hesitant members of the group to participate.

5. Be careful not to dominate the discussion. We are sometimes so eager to express our thoughts that we leave too little opportunity for others to respond. By all means participate! But allow others to also.

6. Expect God to teach you through the passage being discussed and through the other members of the group. Pray that you will have an enjoyable and profitable time together, but also that as a result of the study, you will find ways that you can take action individually and/or as a group.

7. If you are the discussion leader, you will find additional suggestions at the back of the guide.

God bless you in your adventure of love.

Steve Hayner
Gordon Aeschliman

Introducing People and Technology

The image of a leaping ballerina glides across the television screen while a deep, soothing voice tells us: "General Electric brings new things to life." The same company is one of the largest weapons designers in the world.

A major telephone company tells us that its technology will enable us to "reach out and touch someone."

Preachers invite us to join their fellowship by switching to channel 4 on Sunday mornings.

Is technology leading to new definitions of personhood? Is it drawing us away from the mysteries of human relationships? Or is it taking us deeper into those mysteries?

In Hong Kong, journalists debate the negative influence of the electronic media in the tragedy of Tiananmen Square and its aftermath—democracy banners in English played to the immediate presence of the Western media, and people interviewed on television were arrested by the security police. Meanwhile, in Eastern Europe, citizens celebrate the support given by the electronic medium in broadcasting their region's liberation revolution around the globe.

And those images of courage and brutality, beamed across the globe to our television sets, are too easily dulled through their juxtaposition to the daily TV dramas of actor-heroes who divorce themselves from any moral judgment of the violence they depict in the name of entertainment.

Technology helps us manage our complex information needs and aids our ability to get more done in certain disciplines. Yet at the same time some of us sense that our lives are being run by the technology we have created to serve us.

In a court of law a family seeks permission to turn off the machine that has kept their comatose daughter hospitalized and alive on technological support for more than a decade. In the same hospital, an elderly woman receives a heart transplant and renewed life. And in societies where basic community health education is desperately needed for millions of people living with disease and easily rectifiable unsanitary conditions, hospitals import expensive medical equipment for use in treatment of the privileged few.

In the face of innumerable situations like these, questions arise: What is life? What does it mean to have a "quality" life? In a world of high technology and increasing poverty, who decides where limited financial resources will be spent for the benefit of the community?

Are we making choices, or are we being driven by highly complex information systems to produce more with greater efficiency? Are we more than simply consumers of the commodities technology can produce? Is technology influencing us to see ourselves as simply complex machines? Does technological innovation imprison us, so that questions of personhood, community and ethics are ignored in favor of efficiency, progress, innovation, control and production?

What has brought about a society where people gain their personal identity with reference to what they do? or the possessions they own? If our highest values are efficiency, expertise, control of the environment, and growth in the gross national product, do we judge as less human the persons and societies that are less efficient and less expert, and have a lower GNP? What qualities should define personhood, community, and the ethical framework of a society?

As we move from an industrial into an information-based society, a new technological age, we must consider what the ethical vision of such a community will be. We must ask how truth, redemption, freedom and justice will be defined in an age of new information and communication technologies.

The Marks of a Technological Society

First, our society is spending less time in face-to-face communication. More time is being spent with machines—from computer games to television, from information systems to the electronic classroom. As memos, telephone calls, and computer messages replace direct communication, the mystery of being with another person is lost. The simple, natural enjoyment of sitting on the curb and chatting with other members of the "village" is replaced by the draw of the competitive programming of the major networks.

The electronic office so minimizes the need for human contact that John Naisbitt brags in his book *Megatrends 2000* that he didn't have to leave his private mountain chalet while researching the major trends that, in his opinion, will impact the globe during the decade of the nineties.

A wide variety of information systems—print media, television, computer systems—allow people to gain information with limited interaction. The effect is often that they reach conclusions about the lives of people in the absence of genuine dialog with other persons.

Second, new societal divisions are occurring as technology enables some to be information-rich, thereby controlling societal change, while others remain information-poor. Decisions are made on the basis of information expertise, with efficiency being regarded as one of the highest virtues. More and more people are mere cogs in the machinery of modern industry implementing plans that have been set by those in control of information.

Finally, entertainment and news—the soap opera and the "thirty-second news sound bite"—have a fast pace with little time allowed for the viewer to consider the deeper meaning. Action becomes reaction, and reflection is bypassed. The impact of this rapid-fire information can be devastating on management. In such a world, the complexity of ethical analysis is replaced by purely pragmatic concerns. Once a system is established, the means to reach a predetermined end are seldom questioned. Important decisions are made on the basis of briefings that summarize only the benefits to the system. An organization can become so dependent on the efficient machinery it created to help reach certain noble goals that eventually little priority is given to serious reflection on the ethical health of the structure.

Technology is incapable of assisting us in the moral dimension of our lives and ministries. In fact, at times technology has to be bypassed for the sake of moral living.

Applying Scripture

Christians need to apply a biblical understanding as they consider how to respond to the information age. The narratives of creation, the broken reality emerging from the Fall, redemption and reconciliation in Christ, and the coming age are crucial to this understanding.

At creation human beings, female and male, were charged with developing the earth to the glory of God and as a blessing for all. Such a responsibility required the forming of tools—the initial technological innovation. The creation mandate, then, leads us to a positive view of technology as a means for blessing.

With the Fall, reality was distorted by the intrusion of sin. No aspect of life remained unaffected. Subsequently, in the building of the Tower of Babel human beings employed their technological capability in an act of arrogance before God.

Ultimately, it is the narratives of Jesus' life, death and resurrection that reflect what it is to be truly human. The coming age of his reign is depicted as the time when the tools of power—swords and spears—

will be turned into plowshares—implements for bringing blessing from the created world.

If we fail to recognize the Fall as a catastrophic event that affected all subsequent history, we will not understand the history of technological development. Instead of humans' relationship with God being seen as primary, with relationships to other persons and to the created world seen as secondary and derivative, the Fall has led to arrogance: humans idolizing their own skills and capabilities. This is evidenced in the story of Babel when they built a tower to "make a name" for themselves (Gen 11:4).

When technology is viewed as the solution to the basic human dilemma, when technological knowledge is considered a key source of society's "good," it becomes an idolatry. This idolatry twists our understanding of ourselves. We are reduced to highly complex machines, and the mystery of persons within relationships reflecting the Creator God is lost. Rather than blessing the Creator, other persons, and nature, we are subject to dehumanization.

A Christian Response
Probing the Scriptures can help us to see not only how we have come to make an idol of technology, but also how a proper view of persons and tools can be restored. Through Jesus' act of redemption, all idols are cast down at the cross and the empty tomb. No longer is technology to dehumanize the community of God's people. Rather, Christians are to restore it as a means of service.

It is not what we do that defines our humanity. It is not people's technological achievements that will "make a name" for them, but the mysteries of faith, hope and love—the greatest being love. To act while neglecting these three virtues is to fall back into idolatry. In the post-industrial West, where we daily define our personhood in terms of our work, we constantly reflect the idolatry of Babel.

Little wonder that we do not know how to relate to the aged or the disabled. We have failed to perceive that the Bible defines our primary identity not in terms of our work but in our relationships of commitment and forgiveness within the community of faith.

Within a world bent by the Fall, God's people are to confront the distortions brought about by the idolization of technology. As Jacques Ellul emphasizes, it is not plans, strategies or techniques that are to mark the people of God, but a vision of God's shalom, a community marked by love.

As Christians, our role is both prophetic and redemptive. As a prophetic community, we are to confront society's failure to honor the integrity of personhood. As a redemptive community, we are to develop the virtues of character that flow from love. In such a community, efficiency must always be subservient to the aims of love.

Our dilemma in the modern church is that we have lost our sense of belonging to the community of God's people. We tend to perceive our faith as being private—between God and me—for we have failed to understand that God's revelation has always dealt with individuals within the matrix of relationships.

Our world is one of technological complexity. The question of what it means to reflect the image of God confronts us daily. Often the answer leaves us with a sense of ambiguity and pain because of the gulf between our treatment of people under the pressure to uphold efficiency and productivity in a technological society and our understanding of how we really ought to love others. We are tempted to stay away from the community of God's people, but we must not. The dignity of personhood must remain a primary concern for us. That will come to pass as we are informed by a biblical understanding and reformed through fellowship with God and his people.

Elward Ellis, president of Destiny International, has said: "We will not be able to walk in integrity until we realize that mission is not just a way of placing ourselves to respond to need. Rather, mission is a way of participating with God in our discipleship." If we are to effectively confront the complex ethical issues that arise in a technological society, it will flow from a profound reverence for the image of God in all people and the central role of relationships (community) in the daily living out of the gospel.

Caring for God's Creation

Too often we define "being a Christian" only in terms of salvation from our individual sins through the death and resurrection of Jesus Christ. Actually, the story of Jesus begins in the first chapters of Genesis. Here we see God dealing with his entire creation and particularly with the peoples of the world, the descendants of Adam and Eve. This account makes clear that, unlike other gods, the One God is not tied to only one nation.

The spreading of the gospel of Jesus Christ to the ends of the earth has its basis in Genesis 1—11, where God is seen in his interaction with persons and societies before he calls Abraham, the father of the Jewish nation. These first eleven chapters of Genesis are foundational for the remainder of Scripture—and for any definition of personhood.

1. What are the qualities you associate with God as Creator?

Read Genesis 1—2:3.
2. How many times does the word "good" occur in these verses?

3. What is seen as "good" (vv. 4, 10, 12, 18, 21, 25)?

4. What would you say is the climax of Genesis 1?

5. In the creation of human beings (1:26-27), what is different from earlier aspects of creation?

What do you think this suggests about how God regards humanity?

6. What does Genesis 1:27 tell you about the nature of human persons?

7. What are humans given (1:27-28) that is qualitatively different from God's command to the rest of creation?

8. What are we told about God's creation in Genesis 1:31?

9. In Genesis 1:28-30, what responsibilities and commands are given to the man and woman?

10. *Response:* What implications for your own stewardship of the physical world do you draw from this passage?

For Further Study: Read the rest of Genesis 2, and compare and contrast the two accounts.

Study 2

The Necessity
of Community

A professor opened his Biology 101 course at a state university with the following statement: "Human beings are essentially nothing more than columns of air surrounded by a collection of molecules. The mouth is on one end, the anus on the other."

In Genesis 1 we saw God's personal involvement in the construction of humans and in the breathing of life into that creation. A far cry from the professor's undignified description.

Yet humans are more than personally constructed wonders of God. They are related to one another. The essential nature of community is the focus of this study.

1. How do you normally define yourself? (Consider, for example, how we customarily introduce ourselves to others.)

Read Genesis 2:4-25.

2. Recall how many times the word *good* occurred in Genesis 1. Verse 18 of Genesis 2 reveals that in this very good creation one thing is "not good"—for man to be alone. Why do you think God says this?

3. What do you think it was like for Adam to name each living creature (vv. 19-20)?

4. How does Adam respond to the creation of the woman (v. 23)?

5. In verses 21-23 how is the woman defined?

6. How is the social dimension of personhood in verse 23 further explained in verse 24?

7. What image is given of the "goodness" of Adam and Eve's relationship in verse 25?

Why didn't Adam and Eve feel shame?

8. What does Genesis 2 teach you about your identity in relation to others?

9. Throughout Scripture, persons are introduced in reference to relationship—for example, in genealogies. How does this contrast with the individualism of Western society?

10. How do you think your culture affects your self-understanding?

How does it affect your understanding of what it means to be a Christian?

11. What elements are necessary for a biblical understanding of community?

What negative influences does a Western person have to overcome in his or her commitment to a development of the community of believers?

12. What does the separation of the church into peer groups do to the experience of community?

What effect does the lack of cultural and ethnic diversity have on community?

13. *Response:* What practical steps could you take to develop a greater sense of community in your church or fellowship?

The Greatest Catastrophe in History

In Genesis 1 and 2 God is shown creating and defining the limitations of human persons. Chapter 3 of Genesis introduces the serpent's savage attack on this identity. In tempting Adam and Eve to accept an alternative to God's presentation of reality, he destroys community. This study focuses on that great Fall.

1. What is it that appears good, pleasing and desirable about the things that tempt you?

Read Genesis 3.

2. How does Satan tempt Eve (vv. 1, 4-5)?

3. In verse 3 how does Eve show that she is already allowing Satan to distort her words?

4. What is the irony of Satan's temptation in verse 5? (Compare it with Gen 1:27.)

5. How do Satan's words in verse 4 differ from what God has said (see Gen 2:17)?

6. Immediately after the woman and man eat, they hide themselves (v. 8). How can we avoid hiding when we have sinned?

7. What does a lack of vulnerability do to relationships?

How does being vulnerable with other persons help us to be honest before God?

8. Notice that the hiding from each other and from God now progresses further. Describe the interaction between Adam and God (vv. 9-12), and between Eve and God (vv. 13).

9. What curses related to work does God place upon his creation (vv. 17-19)?

Compare these curses with Genesis 1:28-29 and 2:15-16. How is the essence of work changed with the curse?

10. How has the Fall influenced our understanding of work today?

11. In modern society we are too often driven by our professions. Why do you think we define ourselves so much in terms of our jobs?

12. What do you think is the biblical basis for self-understanding?

How do you need to change in the ways you understand yourself?

13. *Response:* Is there someone in your community with whom you need to be reconciled? Commit to doing that now.

For Further Study: Review Genesis 1—2 looking for ways in which God specifically defines or limits people.

Study 4

Unity in the Spirit

Those things that are different about us are what often draws us apart. Yet in God's design, they are the very things intended to bring us together. The New Testament church is often held up as having fully understood community. Meeting daily together, holding all things in common, no one needy among them—these phrases describe much of that early church's life. But it wasn't always that simple! Much of the New Testament was written to bring peace between warring factions. Our study now focuses on that challenge.

1. Think of a time you have seen someone exercising his or her spiritual gift to build the body of Christ. Describe the results of that action.

Read 1 Corinthians 11:17-34.
2. What does Paul say is going on in the Corinthian church (vv. 17-21)?

3. Why does Paul emphasize the need for unity in the celebration of the Lord's Supper?

What does this demonstrate about the nature of the church?

Read 1 Corinthians 12:12-31.
4. Why does Paul emphasize the diversity of roles for those in the body (vv. 14-21)?

5. Even within Paul's experience there appears to be a tendency to define oneself in terms of function rather than unity in Christ. How do you define your role in the church?

6. Exalting one's own gifts, says Paul, is disobedience. In a world in which technology and expertise are high values, what do you think Paul has to say to us?

Read 1 Corinthians 13.
7. In a world affected by the Fall, persons are separated from each other. With the individualism of Western philosophical influences and the sociological changes of the industrial and technological revolutions, people increasingly define themselves in terms of their ideological framework, their work, their feelings, what they own. Contrast this mindset with the substance of the Christian life as it is described in 1 Corinthians 13.

8. Think of several Christians who have had the greatest influence on your life. What is it about them that has particularly impacted you?

Read John 17:20-23.

9. According to Jesus, what is the truest reflection of God's character in human beings?

10. *Response:* Pray for Christ's community to be real in your life and relationships.

For Further Study: 1 Corinthians 12:1-11.

Study 5

Turning Away from Babel

Will we allow technology and our society's definition of personhood to drive a wedge into our community? It can and does happen almost naturally—without our help.

In John 5:17 Jesus declares that as God is working, he too is working. As we respond to Jesus there will be purpose in our lives and meaning within our relationships. It is the context of relationships with God and other persons that should give meaning to our work. Unfortunately, our culture elevates the possession of things and applauds compulsive work habits.

We need to intervene where God's image in people is being destroyed by our society and structures. These passages encourage us to take the lead in bringing God's kingdom back into our relationships.

1. When has your faith been challenged in the workplace?

How did you resolve the situation?

Read Genesis 11:1-9.
2. Work can be either good or evil. We may work in a context where our faith blossoms or be driven by the task so that our sin increases. How do the builders of Babel carry out their work for evil purposes (v. 4)?

3. How does God counteract their plans (vv. 6-8)?

4. Technology can give the illusion that we can control our environment and even human life. But we become possessed by our possessions and alienated by dehumanized synthetic environments. Consider your working environment. How does it strengthen your understanding of who God is?

How does it weaken that understanding?

Read Psalm 127.
5. What basis is offered in this passage for the worth of our work
(v. 1)?

6. Why is it foolish to work long hours (v. 2)?

7. Children (v. 3) are a gift from God and are a sign of his favor on
those who do worthwhile work. In what ways has God rewarded you
for good work?

8. How does your attitude toward work shape how you view your personal relationships?

9. In what ways is the technological revolution around you destroying true community?

10. *Response:* What the Lord requires is "to act justly and to love mercy and to walk humbly" with him (Mic 6:8). How can you use technology to serve society so that the people of God do justice and love mercy?

Study 6

The Mystery of Grace

T. W. Manson said, "Love does not begin by defining its objects: it discovers them." In this final study we ponder how we can discover God in other persons. And ultimately, we must ask how we can live differently so that God is in our community.

1. What are some of the ways in which we attempt to justify ourselves?

Read Luke 10:25-28.
2. Examine the first question the lawyer raises (v. 25). What does the

lawyer reveal about himself in asking this question of Jesus?

3. Why do you think Jesus responds as he does in verse 26?

How does the lawyer answer Jesus' question (v. 27)?

4. What does it mean in your personal life and as a part of the church to love the Lord with all your heart, soul, strength and mind?

5. Consider the disciplines of living out your faith. In what areas do you need encouragement from others?

How can you give and receive greater encouragement?

6. What does it mean to you to love your neighbor as yourself?

What is the chief impediment to this in your life (self-image, lack of time, confused priorities)?

7. How can you improve the sharing of all aspects of your life—worship, relaxation and work—with others?

Read Luke 10:29-35.
8. Why do you think the text says, "He wanted to justify himself" (v. 29)?

What does self-justification do to the development of community?

9. Describe the situation of the person lying at the side of the road.

Think of any people you know who have been brutalized in some way—either physically or psychologically. What is difficult about responding to their needs?

10. Discuss the behavior of the priest and the Levite. Why do we choose to ignore persons in need?

11. Who are the Samaritans in our society? (Suggestions: those working with AIDS patients, those teaching in inner-city public schools.)

12. The Samaritan reflected an attitude of self-giving love. What do we find difficult about expressing this kind of love?

Read Philippians 2:1-18.

13. *Response:* Pray through the passage, asking God to make you corporately more like the community of Jesus.

Suggestions for Leaders

Leading a Bible discussion can be an enjoyable and rewarding experience. But it can also be intimidating—especially if you've never done it before. If this is how you feel, you're in good company. When God asked Moses to lead the Israelites out of Egypt, he replied, "O Lord, please send someone else to do it!" (Ex 4:13). But God's response to all of his servants—including you—is essentially the same: "My grace is sufficient for you" (2 Cor 12:9).

There is another reason you should feel encouraged. Leading a Bible discussion is not difficult if you follow certain guidelines. You don't need to be an expert on the Bible or a trained teacher. The suggestions listed below should enable you to effectively and enjoyably fulfill your role as leader.

Preparing for the Study
1. Ask God to help you understand and apply the passage in your

own life. Unless this happens, you will not be prepared to lead others. Pray too for the various members of the group. Ask God to open your hearts to the message of his Word and motivate you to action.

2. Read the introduction to the entire guide to get an overview of the subject at hand and the issues which will be explored. If you want to do more reading on the topic, check out the resource section at the end of the guide for appropriate books and magazines.

3. As you begin each study, read and reread the assigned Bible passages to familiarize yourself with them. Read the passages suggested for further study as well. This will give you a broader picture of how these issues are discussed throughout Scripture.

4. This study guide is based on the New International Version of the Bible. It will help you and the group if you use this translation as the basis for your study and discussion.

5. Carefully work through each question in the study. Spend time in meditation and reflection as you consider how to respond.

6. Write your thoughts and responses in the space provided in the study guide. This will help you to express your understanding of the passage clearly.

7. It might help you to have a Bible dictionary handy. Use it to look up any unfamiliar words, names or places. (For additional help on how to study a passage, see chapter five of *Leading Bible Discussions,* IVP.)

8. Take the response portion of each study seriously. Consider what this means for your life—what changes you might need to make in your lifestyle and/or actions you need to take in the world. Remember that the group will follow your lead in responding to the studies.

Leading the Study

1. Begin the study on time. Open with prayer, asking God to help the group to understand and apply the passage.

2. Be sure that everyone in your group has a study guide. Encourage the group to prepare beforehand for each discussion by reading the introduction to the guide and by working through the questions in the study.

3. At the beginning of your first time together, explain that these studies are meant to be discussions, not lectures. Encourage the members of the group to participate. However, do not put pressure on those who may be hesitant to speak during the first few sessions.

4. Have a group member read the introductory paragraph at the beginning of the discussion. This will orient the group to the topic of the study.

5. Have a group member read aloud the passage to be studied. (When there is more than one passage, the Scripture is divided up throughout the study so that you won't have to keep several passages in mind at the same time.)

6. As you ask the questions, keep in mind that they are designed to be used just as they are written. You may simply read them aloud. Or you may prefer to express them in your own words. There may be times when it is appropriate to deviate from the study guide. For example, a question may already have been answered. If so, move on to the next question. Or someone may raise an important question not covered in the guide. Take time to discuss it, but try to keep the group from going off on tangents.

7. Avoid answering your own questions. If necessary, repeat or rephrase them until they are clearly understood. An eager group quickly becomes passive and silent if they think the leader will do most of the talking.

8. Don't be afraid of silence. People may need time to think about the question before formulating their answers.

9. Don't be content with just one answer. Ask, "What do the rest of you think?" or "Anything else?" until several people have given answers to the question.

10. Acknowledge all contributions. Try to be affirming whenever possible. Never reject an answer. If it is clearly off-base, ask, "Which verse led you to that conclusion?" or again, "What do the rest of you think?"

11. Don't expect every answer to be addressed to you, even though this will probably happen at first. As group members become more at ease, they will begin to truly interact with each other. This is one sign of healthy discussion.

12. Don't be afraid of controversy. It can be very stimulating. If you don't resolve an issue completely, don't be frustrated. Move on and keep it in mind for later. A subsequent study may solve the problem.

13. Periodically summarize what the group has said about the passage. This helps to draw together the various ideas mentioned and gives continuity to the study. But don't preach.

14. Don't skip over the response question. Be willing to get things started by describing how you have been convicted by the study and what action you'd like to take. Consider doing a service project as a group in response to what you're learning from the studies. Alternately, hold one another accountable to get involved in some kind of active service.

15. Conclude your time together with conversational prayer. Ask for God's help in following through on the commitments you've made.

16. End on time.

Many more suggestions and helps are found in *Small Group Leaders' Handbook* and *Good Things Come in Small Groups* (both from IVP). Reading through one of these books would be worth your time.

Resources

Bell, Daniel. "The Social Framework of the Information Society." *The IT Revolution,* ed. Tom Forester. New York: Basil Blackwell, 1985.

Ellul, Jacques. *The Technological Society.* New York: Vintage, 1964.

Forester, Tom. *High Tech Society.* New York: Basil Blackwell, 1987.

Lyon, David. *The Information Society: Issues and Illusions.* New York: Basil Blackwell, 1988.

Lyon, David. "Modes of Production and Information." *Christian Scholar's Review* 23,3 (1989).

Lyon, David. *The Silicon Society.* Grand Rapids: Eerdmans, 1987.

Lyon, David. "Tubal-Cain and High-Tech." *Christian Arena,* March 1987.

Monsma, Stephen, ed. *Responsible Technology.* Grand Rapids: Eerdmans, 1986.

Roszak, Theodore, *The Cult of Information.* New York: Pantheon, 1986.

Schuurman, Egbert. *Technology and the Future.* Pittsburgh, Penn.: Radix Books, 1980.

World Christian magazine. P.O. Box 40010, Pasadena, Calif., 91104.

Wright, Christopher. *An Eye for an Eye.* Downers Grove, Ill.: Inter-Varsity Press, 1983.